GOD IS ELECTRIC
GOD IS MAGNETIC
GOD IS +VE
GOD IS -VE

A Scientific Theory

SHEILA S. BER

GOD IS ELECTRIC
GOD IS MAGNETIC
GOD IS +VE
GOD IS -VE

A Scientific Theory

SHEILA S. BER

GOD IS ELECTRIC GOD IS MAGNETIC GOD IS +VE GOD IS –VE
A SCIENTIFIC THEORY

iUniverse books may be ordered through booksellers or by contacting:

iUniverse
1663 Liberty Drive
Bloomington, IN 47403
www.iuniverse.com
1-800-Authors (1-800-288-4677)

Because of the dynamic nature of the Internet, any web addresses or links contained in this book may have changed since publication and may no longer be valid. The views expressed in this work are solely those of the author and do not necessarily reflect the views of the publisher, and the publisher hereby disclaims any responsibility for them.

Any people depicted in stock imagery provided by Getty Images are models, and such images are being used for illustrative purposes only. Certain stock imagery © Getty Images.

ISBN: 978-1-5320-6209-4 (sc)
ISBN: 978-1-5320-6211-7 (hc)
ISBN: 978-1-5320-6210-0 (e)

Library of Congress Control Number: 2018913298

Print information available on the last page.

iUniverse rev. date: 12/05/2018

Dedication

This book is dedicated to my dear sons Philip and Bernard Ber, whom I love very much, and to my late husband Henry Ber, whose memory shall never be forgotten. I also dedicate this book to my dear mother Rebecca, with love and appreciation.

To the reader:

The theories in my book are not part of a thesis, and therefore do not require research supporting documentation. They are my own, and have developed over years of self-observation, self-knowledge, curiosity, self-conclusion, and having had a scientific background. No other person has contributed to my theories.

I hope that you have as much curiosity and interest, in the reading of this book as I've had in the writing.

About the Book

This is a scientific theory by Sheila Shulla Ber. An inspirational book for believers and nonbelievers of God and a scientific theory that is educational and easy to discern for readers of all levels.

Reviews by Washington Education Association:

"Complete information! It's this kind of very good read. I have read through and I also am confident that I will study once more later on. You will like just how the author wrote this pdf" (Prof. Darien Mayer).

"I actually started looking over this ebook. It is actually loaded with knowledge and wisdom. It's been printed in an extremely easy way, and just soon after I finished reading through this publication, it basically changed me, changed the way I believe" (Mr. Kristoffer Spinka).

"This book is fantastic. It can be written in basic phrases rather than confusing ones. Your way of life will likely be converted the instant you complete reading this ebook" (Laurie Pouros II).

God: Who really are you? Where are you?

God: Are you among us, or are you one of us?

These are questions we often ask.

God is invisible. God's invisible presence is supremely powerful.

We question also whether God is really powerful, and how?

Can God identify with our suffering, or solve our problems?

We all feel every now and then, that we are anchored in a situation that we have absolutely no control of, and some unexplainable powerful force is at work.

This invisible force can have many different effects on all of us. These effects can be either positive, negative, neutral, miraculous, or disastrous, all at varying degrees.

How can an invisible force be so powerful?

Let's examine this Godly force from a scientific and more realistic point of view.

The main cosmic forces of nature in our universe are:

1. Force of Gravity,

2. Electromagnetic,

3. Electrostatic,

4. Weak nuclear

5. Strong nuclear.

These forces are invisible and have the ultimate control over all bodies, microorganisms, plants, and objects on the planet, or on any planet within the universe.

It is not a coincidence that astrologists have been studying the cosmic planets and stars' movements, and their effect on our planet, for thousands of years.

Astrologists feel very strongly about the ongoing cosmic influence exerted on our planet.

Over time a considerable amount of evidence has been accumulated, suggesting that such phenomenon is real.

My theory reflects the belief that God is either cosmic, earthly, or both.

If God is cosmic, I perceive God therefore, to represent all cosmic entities, to be one of the forces of nature, or perhaps God is the sum of all the cosmic entities and the forces. I'm interested particularly at the power of the well familiar force: **The electromagnetic force.**

The electromagnetic force is magical and powerful, and it can have variations in its power and strength.

Why and how God represents the electromagnetic force?

I believe that God is not only cosmic, but is among us, and resides also in all of us earthly beings.

The cosmic electromagnetic force is connected to all the electromagnetic fields that surround humans, animals, microorganisms, plants, as well as all objects within the universe.

Hence God is the force exuding electromagnetic **energy** that connects all energy fields of each body, plant, or object on this planet, within the universe.

The intensity and strength of each electromagnetic field is relative, whether it is cosmic or earthly.

If the presence of God is everywhere, it is because the electromagnetic force is found everywhere.

Furthermore, if such force connects all beings, in the universe, hence God, who represents such force, would without question, be accessible to every one of us, whether we are believers or non-believers.

DESTINY or FAITH

Humans have learned over time, that they have no complete control over their faith or destiny.

Why?

The answer is: all bodies, plants, microorganisms, and objects on the planet, are controlled or influenced to a great degree by the cosmic forces.

Planets and stars in our galaxy, also exert their influence on all that exists in the universe, whether it is cosmic or earthly.

Every being maneuvers own way in a different and unique manner, according to their chemical makeup, and also the electromagnetic energy of varying strength, that each possesses. All these are combined with the influence of the cosmic electromagnetic force, and the degree of interaction with it.

Faith is determined, not only by each of the individual's body electrochemical makeup, but also by their personal or non personal decision making, and their varied life circumstances.

PRAYERS vs. GOD

The power of prayers!

The question often arises: Why do we pray? How does a prayer help us?

Humans have generally a highly developed brain, that is also active and creative, and it seeks to know and to believe in a super power, or in a higher spiritual being.

We find it more convenient to rely on, and believe in something that is distant, unattainable, or invisible. The invisible that is supernatural, mystique, and is of a higher power.

To exist, we must believe, in order to feel secure, protected, and sheltered from our enemies, predators, nature devastations and other misfortunes of life.

Different cultures have different belief systems. Such systems have existed historically for thousands of years.

The belief system is a very necessary and integral tool for humans' survival, and one of its expressions is carried out by doing prayers.

The reason being, is that when a person believes in something, it is natural that he or she will turn to it for help, for understanding and also for forgiveness, in times of troubles and in times of helplessness.

Prayers automatically produce positive feelings of hope, purpose, and anticipation for better things to come. Hope is also the yearning for prayers to be answered, and to be fulfilled.

Why our prayers are sometimes heard, sometimes not?

Prayers = electromagnetic massages = electromagnetic signals of specific frequencies.

Although prayers, across the globe, are directed towards different godly objects or spirits, collectively they are simply electrical signals that are converted to electromagnetic waves.

Again, beings, plants, animals, microorganisms, bodies of waters, and objects, are all connected through electromagnetic fields.

Additionally, each is encompassed by own electromagnetic energy field that is of a different magnitude and strength. Every action and reaction of beings, is actually carried out, by transmitting and receiving electromagnetic signals/massages at various frequencies and magnitudes.

Electromagnetic signals or massages can be either verbal or non verbal, and are directed usually at whom we wish to connect with spiritually.

The human conscious and subconscious is influenced by the electromagnetic energy, whether it's cosmic, or from the surrounding bodies and/or objects. Therefore it responds selectively or non-selectively to electromagnetic signals.

The response of each individual human conscious/ subconscious to others' prayers, is expressed somewhat similarly, but with some variance that is of distinctive characteristic to each individual electrochemical makeup.

Prayers = transmissions and receptions of electric signals converted to electromagnetic waves, on a quantum level

When we pray we transmit electromagnetic waves that are specific characteristically to a prayer. These waves are of certain frequencies that are within a specific bandwidth of the electromagnetic spectrum.

Prayers are unique signals/messages that are different in comparison with other signals/messages that humans transmit.

Messages have each a specific frequency, that is characteristic to each individual message/signal content.

When someone prays, there are other humans that will probably receive the prayer transmission, consciously or subconsciously, <u>near or far</u>.

Those who are on the receiving side of the transmission, may respond consciously or subconsciously, or may not respond, due perhaps to signal interference/s, or disturbance/s from other electromagnetic fields, of relatively greater magnitude.

Although the recipient may not be fully aware of what type of message/s or information they are receiving, and from whom, regardless, their conscious or subconscious may respond automatically to the signal/s that they are receiving.

The response of the recipient conscious/subconscious, may sometimes experience a delayed reaction, and thus may not always be immediate or automatic.

However, an automatic or a delayed response to someone else's signal/message, at a conscious or subconscious level, may actually be a response to someone else's prayer.

Such transmissions and receptions of electromagnetic signals/messages are again electromagnetic waves, occurring at a quantum level of physics and chemistry.

THE NERVOUS SYSTEM AND ELECTRIC SIGNALS

Some of our actions and reactions are conscious and some are performed by our subconscious.

The human nervous system contains approximately 100 billion nerve cells.

The nervous system consists of the central and peripheral nervous systems. The two systems are intertwined.

Information about the environment is acquired through sensory cells that are specialized to respond to a particular external stimulus.

The sensory cell generates an <u>electrical signal</u> in response to the stimulus. The basic signaling unit of the nervous system is the nerve cell, or neuron, which comes in many different shapes, sizes and chemical content.

Neurons generate and transmit electric signals.

Information via electrochemical signals is received on dendrites and passed on thru an axon.

The electrical signal in axons is a brief voltage change called an action potential, or nerve impulse, <u>which can travel long distances</u>, sometimes at high speeds, without changing size or shape.

When an action potential arrives at the ends of the axon, it interacts with up to thousands of neighbouring cells across synapses.

Thus evidently all beings are complex **electrochemical structures.**

The **nervous system** is a biological system that is made of countless of electric charged particles, at the quantum physics and chemistry level. Thanks to these charged particles, the body is able to transmit and receive information.

The **subconscious** mind is about 88% of one's total mental ability. It is all energy, and hence it is comprised of electric charged particles, that are capable of transmitting and receiving electric signals or massages.

There are different electromagnetic signals that are of different frequencies, and therefore have different codes. It is possible for the recipient humans or animals' conscious and/or subconscious, to decode these different signals.

MENTAL TELEPATHY – THE SILENT UNIVERSAL COMMUNICATION

Mental telepathy is a silent form of communication also occurring at a quantum level.

Every atom or molecule in the universe, emits its own unique frequency, and it is involved in communication between other atoms and/or molecules.

Transmission and reception of frequencies that are directed at other humans and/or animals, can evoke biological processes through chemical and electrical impulses. Such communication process occurs at the invisible sub-atomic level, **the quantum level**.

As discussed before, the human brain can transmit and receive silent messages through electromagnetic energy fields. Messages are each specific to their characteristic content.

In mental telepathy, the messages are different in value from "prayer" messages. These messages are characteristically within other specific bandwidth on the electromagnetic spectrum.

Messages associated with mental telepathy, can be transmitted and received, from a very long distance, and their content or meaning, usually have a great degree of familiarity, when received by the recipient.

At a quantum level, each body, plant, microorganism, and/or object in the universe, is comprised of chemicals that break down to electric particles.
Electric charged particles' movement produce magnetic fields, and vice versa.

Electric particles are charged either positively or negatively, and are constantly moving, constantly attracting or repelling other electric charged particles.

The electromagnetic energy of humans, animals, microorganisms, and objects, is interconnected by each individual vibrating energy field. Similarly, neurons of beings are interconnected not only internally, but also externally, to others' neural networks.

In chemical reactions, each atom attracts or repels another atom/s, depending on the net electrical charge that each atom has.

Atoms bear different net charges. Some atoms have net electron/s (-ve) charge, and some have net proton/s (+ve) charge. The opposites will always attract.

The degree of attraction and repellence, also varies in strength or weakness.

Similarly, on a larger scale, electromagnetic fields will attract or repel other electromagnetic fields, depending on the net electric charge that each electric field has.

Wherever there is electricity, ie moving electric charged particles, there are also <u>always</u> magnetic fields produced, to surround these electric charged particles.

Atoms and molecules are the building blocks of **everything** in the **universe.** Therefore there is constant attracting and repelling movements, at a quantum level going on.

At a visible level, similarly, there is the attraction and repellence occurring at all times, between bodies of humans, between humans and animals, between animals and other animals, as well as between humans/animals and objects. They occur at different levels, and at different intensities.

One example is, when two entities attract, we describe them as being "compatible". When the two repel one another, obviously they would be considered "incompatible".

Each and every interaction between electric charged particles in the universe is unique, and together will emit different frequencies. **Frequencies** depend on variables such as **wavelength** and **speed**.

Electromagnetic transmissions take many forms. Among them, are as discussed earlier, the phenomena, one of which is "prayers", and another is "mental telepathy".

Humans, animals, microorganisms, plants, bodies of water, and objects, are all influenced by each other's electromagnetic energy field. They are also influenced to a great degree by the cosmic electromagnetic energy field/s.

<u>CONCLUSION</u>

My theory suggests that God is the sum of all the cosmic forces and entities in the universe, combined with the collective electrochemical systems which make up humans, animals, plants, microorganisms, bodies of water, and objects. In simple words, God is the sum of all electric charged particles, occurring at a quantum physics and quantum chemistry level, within the universe.

Therefore:

GOD IS THE UNIVERSE ULTIMATE ELECTRICALLY CHARGED PARTICLE. **It is positive, it is negative, or it is neutral, when opposite particles combine, each opposite bearing equal number of charge or charges.**

In seeking the truth about God, I attempted to be as realistic as possible, by examining many facts in science, that may convincingly explain, the invisible presence of God.

God is the ultimate electrically charged particle, or particles in the universe, and all beings are also the product of electric charged particles. Therefore, if I believe in myself, hence I believe in God.
Moreover, if I love and respect myself, I subsequently love and respect God.
God's greatness is our greatness!

I hope that you'll find my scientific theory about God interesting and enlightening.

*Similar to many scientific theories, unfortunately some aspects in my theory are difficult to prove, and I accept that.

KEYWORDS:

Eelectro-magnetic force, God, prayers, universe, mental telepathy, conscious, subconscious, electric charged particles, Quantum physics, quantum chemistry, sub-atomic level, molecular level, electric signals, humans.

Disclaimer.

BIOGRAPHY

I majored in Science in 1991 from the University of Toronto.
Physics and Chemistry were especially, my favourite subjects.

I worked in Microbiology and Chemistry, for about 12 years, in the Pharmaceutical, cosmetics, and toiletry industries.
I was involved in Research & Development, analysis, and formulations of large variety of products.

I am unconventional, but at the same time I like things straight, simple, and uncomplicated.

I have a tendency to analyze everything to extremes. Of course this has its positives, but can have its negatives as well.

I like helping people. I view people, things, and situations, from different perspectives, and try to remain in a neutral position.

Our present digital world is somewhat intimidating, but is rather promising at the same time. It is best to exercise the right balance in our lives.

* * *

SHEILA (SHULLA) BER

Reviews for my book appear at this link:

https://prg.washington.edu/kecc0ri6ix64/30-ralph-goyette-dvm/read-9781508834144-god-is-electric-god-is-magnetic-god-is-ve-god-is.pdf

The book: **Neural activity potential in dead bones theory,** is now on sale at:

www.amazon.com

www.kobobooks.com

www.kindle.com

www.chapters.indigo.ca

www.ebay.com

www.BarnesandNoble.com

God is electric God is magnetic
Book - Sheila Shulla Ber

https://www.youtube.com/watch?v=l37ZNtNxxcg

The above is a musical composition I composed and played, in demonstration of my book God is electric God is magnetic.

www.ingramcontent.com/pod-product-compliance
Lightning Source LLC
Chambersburg PA
CBHW021016180526
45163CB00005B/1973